Getting Started with chipKIT™
The Arduino Compatible PIC32 Based Module

by Chuck Hellebuyck

The publisher offers special discounts on bulk orders of this book.

For information contact:

Electronic Products
P.O. Box 251
Milford, MI 48381
www.elproducts.com
chuck@elproducts.com

The Microchip name and logo, MPLAB® and PIC® are registered trademarks of Microchip Technology Inc. in the U.S.A. and other countries. The Digilent name and logo are registered trademarks of Digilent Incorporated.
 chipKIT™ is a trademark of Microchip Technology Inc.
UNO32™ and MAX32™ are a trademark of Digilent Inc.
All other trademarks mentioned herein are the property of their respective companies.

"Getting Started with chipKIT" is an independent book and is not affiliated with, nor has it been authorized, sponsored, or otherwise approved by Microchip.

Printed in the United States of America
Cover Design by Rich Scherlitz – www.rawmicro.com

Table of Contents

Introduction

New electronic based products are released daily. At the heart of these electronic devices is typically a microcontroller. Over the years many improvements in the tools needed to develop microcontroller based designs have been released. One of the more popular development modules is the Arduino that is so easy to program, artists, designers, hobbyists and even basement engineers have created very interesting projects using this development path. The design is open sourced so many variations of the Arduino exist.

Microchip® Technology Inc. and Digilent® Inc. with software help from Mark Sproul and Rick Anderson of the FUBAR Hackerspace worked together and created a Microchip PIC® based Arduino compatible module called the chipKIT. chipKIT comes in two versions; the smaller chipKIT UNO32 and the larger chipKIT MAX32. These modules are based on the Microchip PIC32 microcontroller and offer many more features than the typical Arduino module. Despite these added features, the chipKIT platform is designed to work with the same simplified programming language used by Arduino and also uses the same type of bootloader programming through a USB connection. The chipkit makes it easy for anyone from beginner to experienced professional to develop interesting projects and products.

In this book I'll show you how to get started with the chipKIT UNO32 using some very simple example sketches (sketch is a software program in the Arduino world) that demonstrate how to use digital inputs/outputs and analog inputs/outputs. From these simple examples you'll have the building blocks to get your electronic project, gadget or product up and running quickly and easily. If you have any questions regarding the content of this book you can contact me via email at chuck@elproducts.com.

Chapter 1 – What is chipKIT

You can purchase many different microcontroller based modules for electronic development but none as powerful and still as easy to use as the chipKIT modules. chipKIT uses a simplified form of C language based around the popular Arduino module. This C compiler includes many pre-written functions to make it easier for the beginner to use. This accomplishes two tasks; it allows complete beginners to get started and it also teaches them the fundamentals of the C language that is used in industry.

The chipKIT in many ways was developed for the artist or mechanical hacker crowds who aren't your normal electronics users. It's proven by the fact that software written for the chipKIT is called a sketch (like an artist's drawing) rather than a program which is the typical name used for software. The chipKIT platform is all open source so all the electrical schematics, circuit board layouts and software are all available to the end user. There are two versions of the chipKIT board; the UNO32 and the MAX32. In this book I'll use the chipKIT UNO32. To better understand the chipKIT UNO32 let me detail what makes up the module.

chipKIT Overview

The chipKIT UNO32 is a microcontroller board based on the PIC32MX320F128H microcontroller. The larger chipKIT MAX32 is based on the PIC32MX795F512L. They both offer Arduino shield (expansion boards are called shields) compatible connections which include 14 digital input/output pins (of which 5 can be used as PWM outputs), 6 analog inputs, and a power header with Reset, 5v, 3.3v, Vin and Ground connections. The chipKIT modules include a USB connection, a power jack, an In Circuit Serial Programming (ICSP) header, and a reset button. The chipKIT modules also offer many extra input/output (I/O) connections beyond the basic Arduino connections but for this book I'll focus on the Arduino/chipKIT common connections.

The chipKIT software is designed to be completely compatible with Arduino. The same functions that control the Arduino can control chipKIT. This means all the Arduino sample files can be used with the chipKIT. This also means the projects in this book will work with Arduino.

The setup for chipKIT is the same as the setup for Arduino. All you have to do is download the software and connect the chipKIT module to a computer with a USB cable to get started. The chipKIT modules can be powered from the USB cable or separately from an AC-to-DC adapter or battery. To program the board you need the Arduino/chipKIT programming software which I'll explain in a few pages. The UNO32and MAX32 are shown in Figure 1-1.

chipKIT™ Uno32™
(Part # TDGL002)

chipKIT™ Max32™
(Part # TDGL003)

Figure 1-1: chipKIT UNO32 and chipKIT MAX32

The PIC32 microcontrollers run at 3.3v but many of the digital I/O are 5v tolerant. All the Arduino compatible pins are 5v tolerant. The Analog pins are 3.3v as well but have protection if 5v is applied. The PIC32 also runs 80 MIPS and has a total of 42 I/O on the UNO32 and 83 I/O on the MAX32.

Module Details

The supporting documentation and details can be found at:

www.microchip.com/chipkit

or

www.digilentinc.com/chipkit

Module Features

Feature	chipKIT MAX32	chipKIT UNO32
Core	80 Mhz, 3.3v, 32-bit	80Mhz,3.3v,32-bit
Flash Memory (KB)	512	128
RAM (KB)	128	16
USB	YES (FS Device/Host, OTG)	-
CAN	2	-
Ethernet	YES	-
DMA	YES	-
PMP/PSP	YES	YES
RTCC	YES	YES
Timers	16/32-bit	16/32-bit
PWM	16/32-bit	16/32-bit
ADC	16 ch. 1Msps, 10-bit	16 ch. 1Msps, 10-bit
Comparators	2	2
I2C	5	2
SPI	4	2
UART	6	2

Power

The chipKIT module can be powered via the USB connection or with an external power supply. The power source is selected automatically by sensing circuitry within the chipKIT module. The module has a 5v regulator and a 3.3v regulator. External power feeds the 5v regulator which then feeds the 3.3v regulator. The external power can come either from a 7-15 volt AC-to-DC adapter with a 2.1mm center-positive plug or a 9 volt battery with a cable that ends in a 2.1mm plug. You can also bypass the 5v regulator and power it direct from 5v with a jumper selection on the circuit board.

Arduino Compatible Headers

The chipKIT UNO32 module was designed to be compatible with the Arduino UNO module even though the chipKIT has dual row connector headers. The UNO32 circuit board dual row connection headers are positioned to accept Arduino compatible shields on the outer row. These outer rows include two 8-pin headers that contain 14 digital I/O, a 6-pin analog header and a 6-pin power header. There are some minor differences so let's go through the header features.

Digital Pins

Each of the 14 Arduino compatible digital pins on the chipKIT UNO32 can be used as inputs or outputs, using pinMode(), digitalWrite(), and digitalRead() functions. Each pin can provide or receive a maximum of 25 mA and has protection so they are 5v compatible. The pins are arranged to match the Arduino UNO with a few minor differences noted below.

Serial: pin0 (RX) and pin1 (TX) – These pins are used to receive (RX) and transmit (TX) serial data and are connected to the corresponding pins of the FTDI USB-to-TTL Serial chip. It's through these pins that the sketch is loaded into memory.

External Interrupts: pin2, (pin 7, 8) - These pins can react to an external signal automatically if setup in software to do so. It can be configured to trigger an interrupt on a low value, a rising or falling edge, or a change in value. The attachInterrupt() function is used to set this up. (Arduino also uses pin 3 as an external interrupt pin but UNO32 has extra external interrupts on pin 7 and 8 instead)

PWM: pin3, pin5, pin6, pin9, pin10 (on UNO32 10 is PWM via jumper JP4) - Provide PWM output with the analogWrite() function. (Arduino has PWM on pin 11 but on UNO32 pin 11 is just digital I/O)

SPI: pin10 (SS via jumper JP4**), pin11 (MOSI), pin12 (MISO), pin13 (SCK)** - These pins support SPI Master communication. (UNO32 allows these pins to be setup as slave pins using jumpers).

LED: pin13 - There is a built-in LED connected to digital pin13. When the pin is HIGH the LED is on, when the pin is LOW it's off.

Analog Pins

The chipKIT UNO32 has 6 analog pins in the same location as the Arduino UNO. There are some minor differences that will be explained here.

Analog to Digital Conversion (ADC): A0 thru A5 - Analog inputs with 10 bit resolution (i.e. 1024 different values). UNO32 analog pins measure from ground to 3.3 volts. The reference voltage can be changed to a lower voltage using the AREF pin and the analogReference() function. (Arduino analog pins read from ground to 5v)

I^{2}C: A4 (SDA), A5 (SCL) – These pins support I^{2}C communication using the Wire library. On UNO32 you need to move two jumpers on the circuit board to use pins A4 and A5 as I2C pins.

Analog Reference: AREF - Reference voltage pin for the analog inputs. Used with analogReference(). This pin is on the digital header row.

Power Pins

The chipKIT UNO32 shares the same power connections as the Arduino UNO. They are described here.

RESET – This pin connects to the microcontrollers reset pin and pulling this pin to ground will reset the UNO32 to run the sketch in its memory from the very beginning.

VIN – This pin his connected to the center tap of the 2.1 mm connector and makes it easy to connect to the input voltage of the UNO32 board.

5V - This pin has the regulated five volt power produced by the UNO32's five volt regulator.

3V3 - A 3.3 volt supply is generated by the on-board 3.3v regulator.

GND – Two ground pins are on this header.

Special Note:
chipKIT modules have many extra digital and analog pins outside of the normal core Arduino connections. These will be considered an advanced topic and will not be covered in full here since this book is dedicated to those just getting started. Check the microchip.com/chipkit or digilentinc.com/chipkit websites for more details on these extra I/O features.

Jumper Settings
The chipKIT UNO32 and chipKIT MAX32 have many options due to its expanded I/O setup. There are several jumpers that make a selection for certain pins on the module. The UNO32 jumpers are explained next.

Note: These are UNO32 settings only (MAX32 are different).

RD4 10 RG9

JP4
RD4 Position – PIN 10 is PWM pin
RG9 Position – PIN 10 is SPI Slave Select (SS) pin

JP5
Master Position – PIN 12 is SPI SDO (MISO)
Slave Position – PIN 12 is SPI SDI (MOSI)

JP7
Master Position – PIN 11 is SPI SDI (MOSI)
Slave Position – PIN 11 is SPI SDO (MISO)

JP6
A4 Position – PIN A4 is Analog Input
RG3 Position – PIN A4 is I2C SDA pin

JP8
A5 Position – PIN A5 is Analog Input
RG2 Position – PIN A5 is I2C SCL pin

JP2

BYP Position – Power at J4 Power Input (also VIN pin) bypasses 5v regulator goes direct in to 3.3v regulator (Max 6v in when in this mode!)

REG Position – Power at J4 Power Input (also VIN pin) goes to 5v Regulator input before 3.3v regulator (Max 15v in when in this mode!)

Communication

The chipKIT UNO32 has a number of options for communicating with a computer, another chipKIT, or other microcontrollers. The chipKIT chip has an internal UART serial communication peripheral which is available on digital pins 0 (RX) and 1 (TX). An FTDI FT232RL chip on the board converts this serial communication to USB. The FTDI drivers (included with the chipKIT software) provide a virtual com port to software on the computer. The chipKIT MPIDE software also includes a serial monitor which allows simple ASCII data to be sent to and from the chipKIT board.

The FTDI chip has RX and TX LEDs on the board that flash when data is being transmitted via the FTDI chip. The Arduino Software Serial library is also available that offers software functions for serial communication on any of the UNO32's digital pins.

The chipKIT software also supports I2C and SPI communication through the chipKIT/Arduino software libraries.

Programming
The chipKIT UNO32 control code is written using the chipKIT programming software running on a PC or MAC (Linux is also supported but not covered in this book) which is also called the chipKIT MPIDE which stands for Multi-Purpose Integrated Design Environment. MPIDE can be used to program chipKIT modules or any Arduino module.

The UNO32's PIC32 microcontroller is programmed using a bootloader or software method so you don't need a separate hardware programmer. The chipKIT software has an editor for writing the sketches and then a built in compiler and bootloader interface. With a single click of the mouse, your software will be compiled into 1's and 0's and then sent to the flash memory of the UNO32.

For advanced users you can also bypass the bootloader and program the microcontroller through the ICSP (In-Circuit Serial Programming) header with an extra hardware programmer.

Automatic (Software) Reset

Besides the physical reset button, the chipKIT UNO32 can be reset by software running on a connected computer. One of the hardware flow control lines (DTR) of the FT232RL is connected to the reset line of the chipKIT chip through a 100 nanofarad capacitor. When this line is pulled low, the reset line drops long enough to reset the chip. The chipKIT software uses this capability to allow you to upload code by simply clicking on the upload icon in the MPIDE.

This setup has other implications. When the UNO32 is connected to either a computer running Windows, Mac OS X or Linux, it resets each time a connection is made to it from software (via USB). When reset, the UNO32 bootloader will run for a half-second or so to look for new code to download. If your application trying to communicate with the UNO32 starts off by sending data bytes, the bootloader may intercept the first few bytes of data sent to the board after a connection is opened. Its best to make sure that the software communicating with the UNO32 waits a few seconds after opening the connection before sending any data.

Physical Characteristics

The UNO32 PCB is about 2.7 and 2.1 inches with the USB connector and power jack extending beyond these dimensions. Three screw holes on the UNO32 align with the Arduino UNO and allow the board to be attached to a surface or case. If you want to make your own stack on board (known as a shield in the Arduino/chipKIT world) then note that the distance between digital pins 7 and 8 is not a standard 0.100" spacing. This non-standard connection layout makes it a little more difficult to create a shield board from a standard 0.100" spacing protoboard. There are offset connectors available that allow you to plug a standard protoboard into an UNO or UNO32 module. There are also many proto-shields available with the required spacing to fit the UNO/UNO32 layout.

Hardware for Projects

The hardware for the projects in this book can all be built on the breadboard. Each chapter project includes a drawing of the setup on a breadboard layout. For those that don't have much hardware experience, the projects were also written to work with the CHIPINO demo shield. You can get this from chipaxe.com or several other sources. Email me (chuck@elproducts.com) or google search for a complete list of sources. More details on this shield is in the appendix of this book.

CHIPINO Demo Shield

How to get chipKIT running on Mac

These are the fundamental steps to get chipKIT running on a MAC:

- Download the chipKIT MPIDE
- Install the USB driver
- Connect the board
- Run the MPIDE
- Create a Blink LED project
- Run the Blink project on a chipKIT module

1 | Download the MPIDE environment

The software can be downloaded by clicking on the MAC OS X link on the software download page at either of the locations below:

www.microchip.com/chipkit
www.digilentinc.com/chipkit

You will have two files, the MPIDE programming software and the FTDI drivers. Slide the MPIDE icon over the applications folder to install the MPIDE programming software.

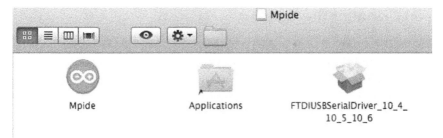

Figure 1-2: chipKIT Files for MAC

2 | Install the USB drivers

Unless you've already used a product with a USB to RS232 FTDI chip, you will need to install the drivers for the FTDI chip on the chipKIT board. Click on the FTDI Drive icon to install the driver. Follow all the steps until installed.

Figure 1-3: Install FTDI Drivers

3 | Connect the board

Connect the USB cable to one of the MACs USB ports and then to the USB connector of your chipKIT board. The chipKIT module uses a mini-A USB cable which is different from the cable Arduino uses.

4 | Run the MPIDE environment

Click on the MPIDE icon to launch the environment and you should see a screen similar to the one in Figure 1-4.

22

Figure 1-4: Arduino/chipKIT Programming Environment

5 | Create a Blink LED project

First select your UNO32 board from the Tools>Board menu. In this book's examples we'll use the UNO32 (Figure 1-5). Then select the serial port you have the chipKIT connected to under the Tools>Serial Port menu (Figure 1-6). On the MAC, this should be something with usbserial in the name. Now you are ready to create your first sketch.

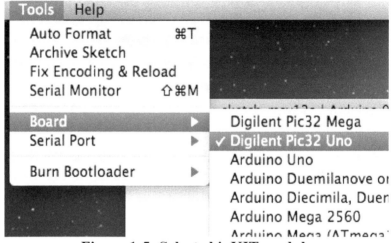

Figure 1-5: Select chipKIT module

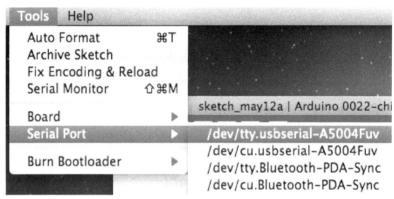

Figure 1-6: Select Serial Port

Many pre-written example sketches are included with the chipKIT environment so it's easier to start with a working example to get your first sketch running. Open the LED blink example sketch: by clicking on the following menu option: File > Sketchbook > Examples > Digital > Blink (shown in Figure 1-7).

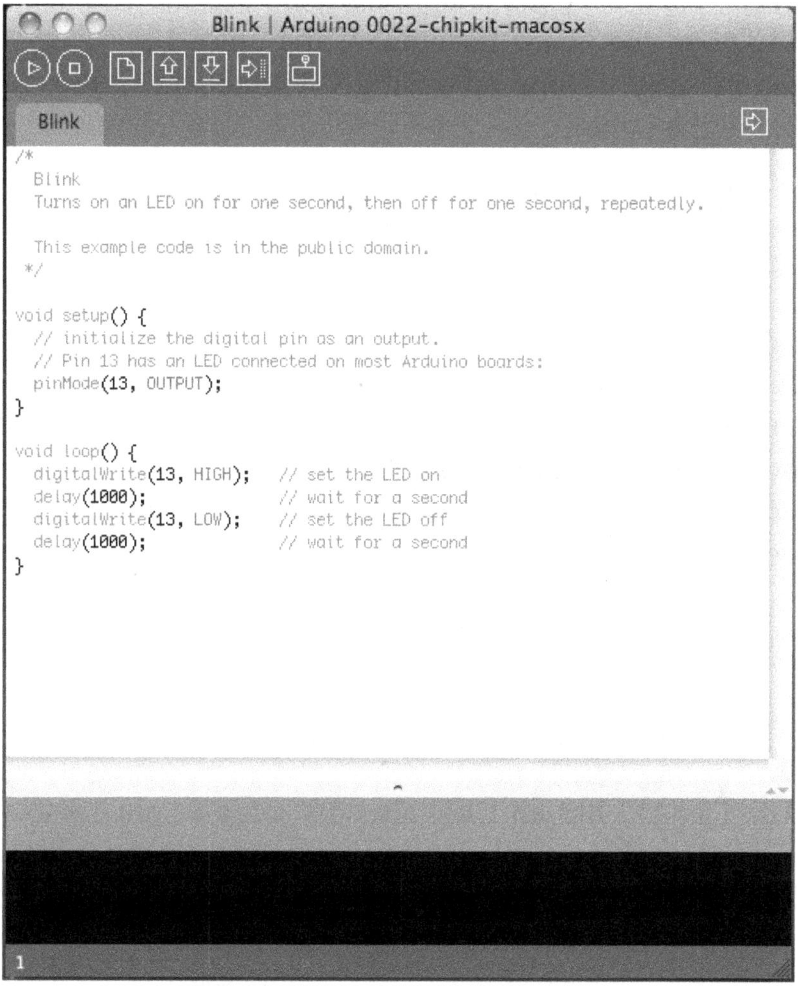
Blink | Arduino 0022-chipkit-macosx
Blink
/*
Blink
Turns on an LED on for one second, then off for one second, repeatedly.
This example code is in the public domain.
*/
void setup() {
// initialize the digital pin as an output.
// Pin 13 has an LED connected on most Arduino boards:
pinMode(13, OUTPUT);
}
void loop() {
digitalWrite(13, HIGH); // set the LED on
delay(1000); // wait for a second
digitalWrite(13, LOW); // set the LED off
delay(1000); // wait for a second
}
1

Figure 1-7: Blink LED Sketch

6 | Run the Blink project on UNO32

Click the "Upload" button in the environment which is the sideways arrow shown highlighted in Figure 1-8. Wait a few seconds and you should see the RX and TX LEDs on the board flashing. If the upload is successful, the message "Done uploading." will appear in the status bar at the bottom.

Figure 1-8: Upload sketch to chipKIT board

A few seconds after the upload finishes, you should see the pin 13 (LD4) LED on the board start to blink. If it does, congratulations! You've successfully programmed your UNO32. If not, go back through the steps to see if you missed something.

Note: For user help visit the chipKIT forum at:

www.chipkit.org/forum

The UNO32 has an LED already wired to pin 13 so you didn't need to connect any circuitry to the chipKIT board. In future projects you will need to connect the proper components to get the sketch to work.

How to get chipKIT running on Windows

These are the fundamental steps to get chipKIT running on a PC:

- Download the MPIDE environment
- Install the USB driver
- Connect the board
- Run the MPIDE
- Create a Blink LED project
- Run the Blink project on a UNO32

1 | Download the MPIDE

The MPIDE programming software can be downloaded by clicking on the windows link on the software download page at either of the locations below:

www.microchip.com/chipkit
www.digilentinc.com/chipkit

It's a .zip file that you unzip to a location on your hard drive. When the download finishes, unzip the downloaded file. You'll end up with a file structure as shown.

drivers
examples
hardware
java
lib
libraries
reference
tools
cygiconv-2.dll
cygwin1.dll
libusb0.dll
mpide
revisions
rxtxSerial.dll

2 | Connect the board

Connect the USB cable to one of the PCs USB ports and then to the USB connector of your chipKIT UNO32 board. The chipKIT module uses a mini-A USB cable which is different than the Arduino.

3 | Install the USB drivers

The installation process will depend on what operating system you are running. The chipKIT modules use an FTDI USB interface chip so it's best to follow the installation instructions at specific to your operating system. The location is at the following address.

http://www.ftdichip.com/Support/Documents/InstallGuides.htm

Here is a typical step by step installation on a PC running XP.

Figure 1-9: FTDI Wizard Screen

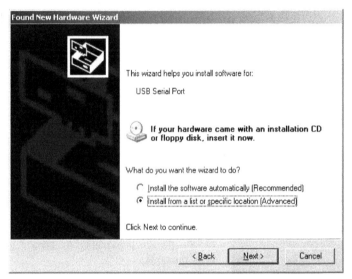

Figure 1-10: Select Specific Location Option

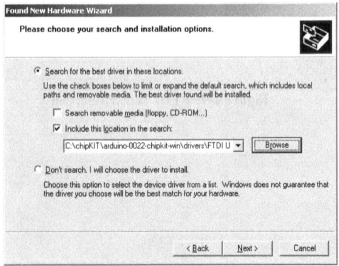

Figure 1-11: Click on Browse to Locate Drivers

30

Figure 1-12: Select the FTDI USB Drivers Folder

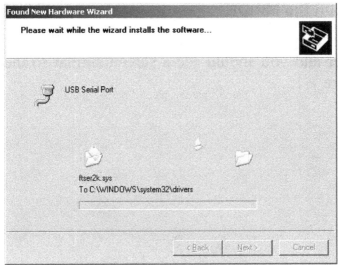

Figure 1-13: Driver Installation

Figure 1-14: FTDI Driver Installation Complete

4 | Run the MPIDE

Click the  mpide icon in the chipKIT environment folder and you should see a screen similar to the one in Figure 1-15.

Figure 1-15: Arduino/chipKIT Environment

5 | Create a Blink LED Project

Select the UNO32 board from the Tools | Board menu.

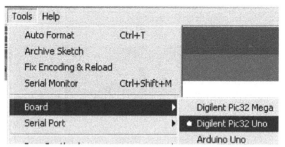

Figure 1-16: Select chipKIT Board

Next, select the proper COM port. To verify, open the Windows Device Manager in the Hardware tab of System control panel.

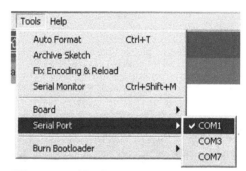

Figure 1-17: Select the COM Port

Open the LED blink example sketch:
File > Sketchbook > Examples > Digital > Blink

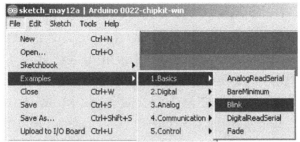

Figure 1-18: Select Blink Sketch

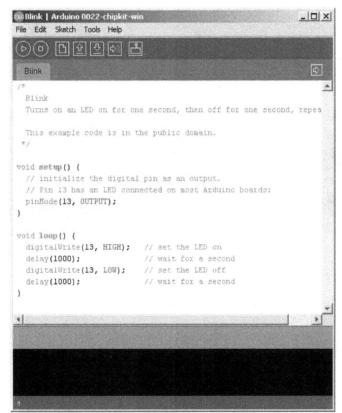

Figure 1-19: Blink Sketch Window

35

6 | Run the Blink project on UNO32

Click the "Upload" button in the environment which is the sideways arrow shown highlighted in Figure 1-20. Wait a few seconds and you should see the RX and TX LEDs on the board flashing. If the upload is successful, the message "Done uploading." will appear in the status bar.

Figure 1-20: Upload sketch to chipKIT board

Figure 1-21: Uploading to chipKIT Board

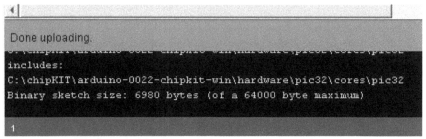

Figure 1-22: Sketch is Loaded in chipKIT

A few seconds after the upload finishes, you should see the pin 13 (LD4) LED on the board start to blink. If it does, congratulations! You've successfully programmed the chipKIT module. If not, go back through the steps to see if you missed something.

Note: For user help visit the chipKIT forum at:

www.chipkit.org/forum

The chipKIT has an LED already wired to pin 13 so you didn't need to connect any circuitry to the chipKIT board. In future projects you will need to connect the proper components to get the sketch to work.

MPIDE

The MPIDE is the same for MAC, Windows and Linux and contains a text editor where you write your chipKIT sketch. A toolbar with buttons for common functions is at the top of the screen along with a series of menus. There is a message area at the bottom to give feedback when saving or exporting a file. It also displays any code errors. The toolbar functions are explained below.

Verify/Compile
Checks your code for errors.

Stop
Stops the serial monitor, or unhighlight other buttons.

New
Creates a new sketch.

Open
Presents a menu of all the sketches in your sketchbook.

Save
Saves your sketch.

Upload to I/O Board
Compiles your code and uploads it to the chipKIT board.

Serial Monitor
Opens the Serial Communication Window.

Additional commands are found within the five menus: File, Edit, Sketch, Tools, Help. The menus are context sensitive which means only those items relevant to the work currently being carried out are available The menu options are as follows:

File
Here you can open existing sketches or create and save new ones.

Edit
Copies the code of your sketch to the clipboard in a forum suitable for posting to a forum, complete with syntax coloring.

Copies the code of your sketch to the clipboard as HTML, suitable for embedding in web pages.

Sketch

Verify/Compile
Checks your sketch for errors.

Import Library
Adds a library to your sketch by inserting #include statements at the code of your code.

Show Sketch Folder
Opens the sketch folder on the desktop.

Add File...
Adds a source file to the sketch (it will be copied from its current location). The new file appears in a new tab in the sketch window. Files can be removed from the sketch using the tab menu.

Tools

Auto Format
This formats your code nicely: i.e. indents it so that opening and closing curly braces line up, and that the statements instead curly braces are indented more.

Board
Select the board that you're using.

Serial Port
This menu contains all the serial devices (real or virtual) on your machine. It should automatically refresh every time you open the top-level tools menu.

Burn Bootloader
The items in this menu allow you to burn a bootloader onto the microcontroller on an chipKIT board. This is not

required for normal use of a chipKIT board but is useful if you want to build your own chipKIT.

Help

This selection allows you to look up help topics that may not be in the manual or this book.

Sketchbook

The chipKIT environment uses a folder to contain all the files in a project. The first time you run the chipKIT software, it will automatically create a folder for your sketch. You can view or change the location of the sketch location from with the Preferences option under the file menu selection.

Tabs, Multiple Files, and Compilation

The environment allows you to break up your sketch into multiple files that get combined into one big file when you are ready to program the chipKIT hardware. The environment has tabs so each file can appears in its own tab. These can be normal chipKIT code files (no extension), C files (.c extension), C++ files (.cpp), or header files (.h).

Uploading

As described in the blink example earlier, to upload your sketch, you need to select the correct items from the

Tools > Board and Tools > Serial Port menus. Once you've selected the correct serial port and board, press the upload button in the toolbar or select the Upload to I/O Board item from the File menu. Current chipKIT boards will reset automatically and begin the upload. You'll see the RX and TX LEDs blink as the sketch is uploaded. The MPIDE will display a message when the upload is complete, or show an error.

Libraries

Libraries are the heart of what makes the chipKIT easy to use with the hardware. The functionality in the microcontroller is easily controlled by these prewritten sections of code you can include in your sketch. To use a library in a sketch, select it from the Sketch > Import Library menu. This will insert one or more #include statements at the top of the sketch and compile the library with your sketch. Because libraries are uploaded to the board with your sketch, they increase the amount of space it takes up. If a sketch no longer needs a library, simply delete its #include statements from the top of your code.

Some libraries are included with the chipKIT software while others can be downloaded from a variety of sources. To install a custom library, create a directory called "libraries" within your sketchbook directory. Then copy the library there. For example, to install the "DateTime" library, its files should be in the /libraries/DateTime sub-folder of your sketchbook folder.

Serial Monitor

The chipKIT module can send back serial data through the same programming connections of the USB cable. The data sent can be displayed on the Serial Monitor built into the chipKIT Environment. To use the Serial Monitor you need to choose the baud rate from the drop-down that matches the rate passed to Serial.begin in your sketch. To send data to the board, enter text and click on the "send" button or press enter.

Preferences

Some preferences can be set in the preferences dialog (found under the chipKIT menu on the Mac, or File on Windows and Linux). The rest can be found in the preferences file, whose location is shown in the preference dialog.

chipKIT C Compiler

The chipKIT uses a programming language based on C. The chipKIT uses names that are not typical to C programming and also has pre-built commands to make it easier for the beginner to get started quicker. The C language gets converted into the 1's and 0's the chipKIT microcontroller needs by the chipKIT C compiler. To use it properly its best to know the basics and this chapter will cover those details.

Sketch

A sketch is the name that chipKIT uses for a program. It's the software that is loaded and run on a chipKIT board. The software below is a sample sketch for flashing the pin 13 LED. I'll describe the various sections as we go through this chapter.

```
/*
Blink
Turns on an LED on for one second, then off for one second, repeatedly.
The circuit:
LED connected from digital pin 13 to ground.

Note: On most chipKIT boards, there is already an LED on the board
connected to pin 13, so you don't need any extra components for this example.
*/

int ledPin = 13;   // LED connected to digital pin 13

// The setup() runs once, when the sketch starts
void setup()
        {
        pinMode(ledPin, OUTPUT); //initialize the digital pin as an output
        }

// The main loop runs over and over again until power is removed
void loop()
        {
        digitalWrite(ledPin, HIGH);   // set the LED on
        delay(1000);              // wait for a second
        digitalWrite(ledPin, LOW);   // set the LED off
        delay(1000);              // wait for a second
        }
```

Comments

The first few lines of the Blink sketch are a comment block:

```
/*
Blink
Turns on an LED on for one second, then off for one second, repeatedly.
The circuit:
LED connected from digital pin 13 to ground.

Note: On most chipKIT boards, there is already an LED on the board
connected to pin 13, so you don't need any extra components for this example.
*/
```

Everything between the /* and */ is considered a comment block and does not get compiled for the chipKIT when it runs the sketch. Comment blocks are used to describe what the sketch does, how it works, or why it's written the way it is. It's a good practice to comment your sketches, and to keep the comments up-to-date when you modify the code.

There's another style for comments. They start with // and continue only until the end of the line.

```
int ledPin = 13;          // LED connected to digital pin 13
```

The section "LED connected to digital pin 13" is a treated as a comment and not compiled into the chipKIT.

Sketch Variables

A variable is a place for storing a piece of data in RAM. It has a name, a type, and a value. For example, the line from the Blink sketch declares a variable with the name

44

"ledPin". It is created as a type "int" and initialized with the value of 13. It's used to indicate which chipKIT pin the LED is connected to. Every time the name ledPin appears in the code, its value will be retrieved. In this case, the person writing the sketch could have chosen not to bother creating the ledPin variable and instead have simply written 13 everywhere but then later if they wanted to change the pin connection they would have to replace it multiple times. Using the ledPin label allows you to change the LED at one spot. Variables are declared before the setup section.

```
int ledPin = 13;    // LED connected to digital pin 13

// The setup() runs once, when the sketch starts
void setup() {
}
```

setup() and loop()

There are two special functions that are a part of every chipKIT sketch: setup() and loop(). The setup() is called once, when the sketch starts. It's a good place to do setup tasks like setting pin modes or initializing libraries. The loop() function is called over and over and is heart of most sketches. You need to include both functions in your sketch, even if you don't need them for anything.

```
void setup() {
  // put your setup code here, to run once:

}
```

```
void loop() {
  // put your main code here, to run repeatedly:

}
```

{ } Curly Braces

Curly braces or curly brackets are important characters in any C programming language which the chipKIT software is based on. An opening curly brace "{" must always be followed by a closing curly brace "}" or you will get an error when you compile. Beginning programmers, and programmers coming to C from the BASIC language often find using braces confusing. After all, the curly braces replace the RETURN statement in a subroutine, the ENDIF statement in a conditional IF statement and the NEXT statement in a FOR loop. So think of curly braces as the beginning and end of any function.

; semicolon

A command line inside your sketch is called a statement in the C language. Each statement needs to mark the end with a semicolon. A statement line can extend multiple lines in your sketch and end with a semi-colon to let the compiler know where to end. The most common error a beginner will see is a missing semi-colon. Two statements will be combined by the compiler and you'll get an unrecognized statement error because of that missing semi-colon.

```
//Simple example of how to end a statement.
int a = 13;
```

const

The chipKIT software offers the const keyword and is the preferred method for defining constants. The const keyword is short for constant and really just a modifier to a variable that makes it read only and cannot be changed at any time in the program.

```
//    constants   won't    change.    Used    here    to
// set pin numbers:

const int ledPin =  12;   // the number of the LED pin
```

#include

#include is used to insert external files that contain definitions of pre-built functions. This is also known as including a library. A library will have a header file or .h file with library. The #include directive is used to include that header file within your sketch. The #include allows you to include both chipKIT libraries and standard C language libraries. The #include in excercised before the compiler is run so the #include line does not end with a semi-colon. It's placed above the setup section.

```
#include "pitches.h"

// notes in the melody:
int melody[] = { NOTE_C4, NOTE_G3,NOTE_G3, NOTE_A3,
NOTE_G3,0, NOTE_B3, NOTE_C4};

// note durations: 4 = quarter note, 8 = eighth note, etc.:
int noteDurations[] = { 4, 8, 8, 4,4,4,4,4 };

void setup() {
```

This covers the basics of programming for the chipKIT but it's plenty of information for you to start creating your own sketches for running on the chipKIT. We'll start by modifying the Blink example and drive an LED from a header pin.

Chapter 2 – Flash an External LED

The first project every beginner needs to do is get an LED to flash. This proves out many things: 1) The hardware is connected properly 2) The software compiles without errors 3) The module can communicate and be programmed from the computer thru a USB connection. 4) We are ready to move on to more complex projects.

If at any point you can't get a project to run on the chipKIT, flashing an LED is always a good indicator that the basic setup is working properly.

Hardware

Connect the chipKIT UNO32 per the drawing in Figure 2-1. The long lead of the LED is connected to pin 12, the short lead is connected to the 220 ohm resistor. The other end of the resistor is connected to the Gnd pin of the chipKIT UNO32

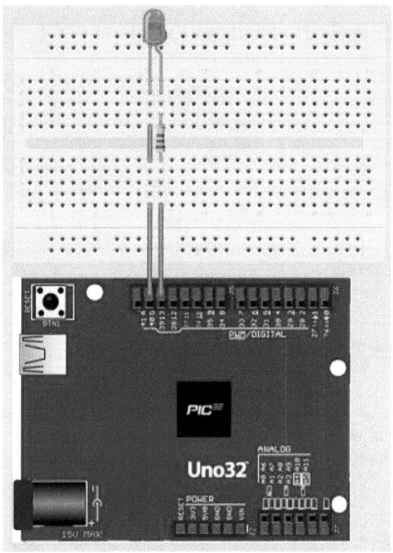

Figure 2-1: Final Flash LED Project

Software

```
/*
  Blink
 Turns on an LED on for one second,
 then off for one second, repeatedly.

*/

void setup() {
  // initialize the digital pin 12 as an output.

  pinMode(12, OUTPUT);
}

void loop() {
  digitalWrite(12, HIGH);    // set the LED on
  delay(1000);               // wait for a second
  digitalWrite(12, LOW);     // set the LED off
  delay(1000);               // wait for a second
}
```

How It Works

The software is quite simple but still requires some explanation.

The top of the software contains a header block that describes what the sketch will do. Any time you want to add a block of text that goes over many lines, just place the text between the /* and */ characters. This indicates to the compiler that the text is just comments and not code.

```
/*
  Blink
 Turns on an LED on for one second,
 then off for one second, repeatedly.

*/
```

51

The first code section of any chipKIT sketch is the setup section where any commands are placed that need to run only once before running the main loop. Typically this will be setting up the digital I/O pins and presetting any variable values. Everything between the curly brackets are part of the setup routine.

```
void setup() {

}
```

Within the setup section we can place comments that are ignored by the compiler but allow us to make notes for future reference. A double backslash before the comment is all you need to add to make any text a comment. You have to keep it in one line though. If the text wraps around to a second line then the /* and */ should be used.

The digital pins need to be setup to either be an input or an output. This is done with the pinMode function. The parameters in the parenthesis select the pin and the direction. We want to control the LED with a high or low output signal from pin 12 so we set pin 12 to an output using the line below.

```
  pinMode(12, OUTPUT);
```

The main sketch loop is the section of code that runs over and over again. All the command lines need to be contained within the curly brackets. The main loop will run continuously as long as power is not removed from the chipKIT or the reset switch is pressed.

```
void loop() {

}
```

The first control is the digitalWrite function that drives pin 12 high which lights the LED.

```
digitalWrite(12, HIGH);    // set the LED on
```

The next line is the delay function. This just creates a one second delay as the value 1000 represents 1000 milliseconds or one second.

```
delay(1000);               // wait for a second
```

The sketch then turns the LED off by setting the same pin low.

```
digitalWrite(12, LOW);     // set the LED off
```

An additional delay function line delays another second.

```
delay(1000);               // wait for a second
```

The sketch then jumps back to the top of the loop since the second curly bracket is encountered. This operation will repeat over and over again to create a simple blinking light.

Next Steps

Simple next steps are to change the pause value to a lower number to flash the LED faster. You could also connect the LED to a different pin and then change the number in the high and low command lines to make that new connection pin flash the LED. You could also control two LEDs at once with a second digitalWrite line. In fact the next project uses three to create a traffic light.

Chapter 3 – Train Crossing

Flashing an LED is a great place to start but lets put it to some practical use. Have you ever stopped at a train crossing and seen the red lights flash back and forth to indicate a train is coming? This can be accomplished real easy with two LEDs. This project can also be built into a model train railroad crossing sign to make it a little more realistic.

Hardware

The hardware is similar to the blink LED project except now there are two LEDs to control. The LEDs are driven on by a high signal on the digital pin that drives them. A low shuts them off. A resistor is placed in series with the LEDs to limit the current. A 1k ohm resistor works fine.

The left LED is connected to the digital pin number 13. The right LED is connected to pin 10. A series resistor is used to limit the current and any value from 220 ohms to 1k ohms can be used. Both LEDs are red on the CHIPINO demo shield.

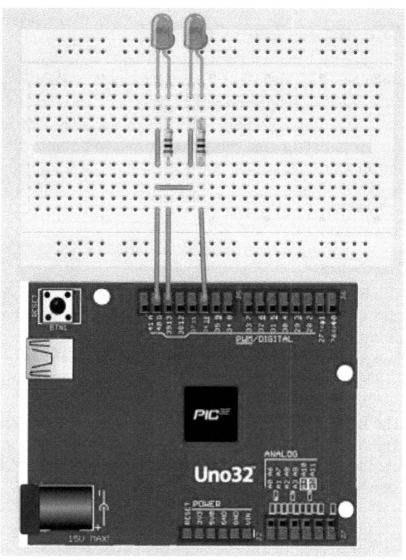

Figure 3-1: Train Crossing Project

```
/*
 Train Crossing
 Blinks two LEDs back and forth like a train
crossing
 */

void setup() {
  pinMode(13, OUTPUT);
  pinMode(10, OUTPUT);
}

void loop() {
  digitalWrite(13, HIGH);// set the left LED on
  digitalWrite(10, LOW); // set the right LED off
  delay(1000);           // wait for a second
  digitalWrite(13, LOW); // set the LED off
  digitalWrite(10, HIGH);// set the right LED on
  delay(1000);           // wait for a second
}
```

How it Works

The first section is just a comment header to describe the sketch.

```
/*
 Train Crossing
 Blinks two LEDs back and forth like a train
crossing
 */
```

The setup loop establishes the two pins to be used as outputs.

```
void setup() {
  pinMode(13, OUTPUT);
  pinMode(10, OUTPUT);
}
```

The main loop controls the two separate LEDs by controlling the digital pins they are connected to. Setting pin 13 to high and pin 10 to low lights the left LED. A delay of 1000 keeps the LED on for 1 second.

```
void loop() {
  digitalWrite(13, HIGH);// set the left LED on
  digitalWrite(10, LOW); // set the right LED off
  delay(1000);           // wait for a second
```

The second section does the opposite and lights the right LED on pin 10 and shuts off the right LED on pin 13. Another delay of 1000 is added.

```
  digitalWrite(13, LOW); // set the LED off
  digitalWrite(10, HIGH);// set the right LED on
  delay(1000);           // wait for a second
}
```

The program stays in a continuous loop alternating the lighting of the LEDs to create the train crossing display.

Next Steps

The delay can be changed to make the LEDs flash faster or slower. As mentioned these could be built into a model train crossing sign to make it realistic looking.

Chapter 4 – LED Traffic Light

A traffic light is something we see daily but with the chipKIT UNO32 we can build our own with a few LEDs and resistors. The trick is to have the UNO32 control each different color LED with a separate digital pin. This way the software can control which LED is on and how long it stays on. Once we have it built then we can control the operation with software.

Hardware

The connections to the LED are similar to the connections in the flash LED example. The long lead of the LED is the positive or anode pin. This is connected to the UNO32 digital pin that will control it. The green LED is connected to pin 12, the yellow LED to pin 11 and the red LED to pin 10. A resistor is in series between the digital pin and the LED anode lead. This is needed to limit the current so the LED doesn't burn-up.

The shorter lead of the LED is the cathode or negative lead. All the LEDs cathodes are connected together through jumper wires. They connect to the ground pin on the same digital pin header. When the digital pin connected to an LED is high, that LED will light. When a digital pin is low, the LED is off.

The hardware connections are shown in Figure 3-1. Red is on the right, yellow in the middle and green on the left.

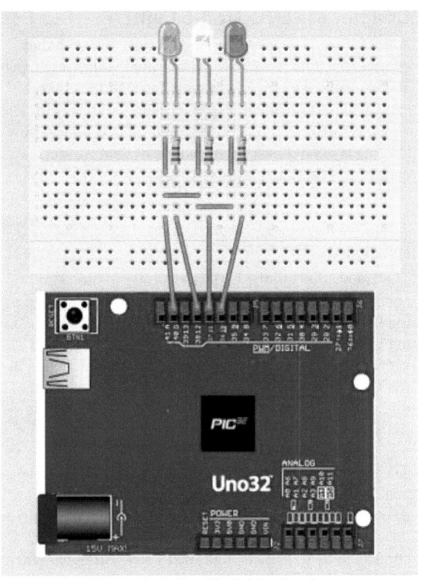

Figure 4-1: Traffic Light Project

Software

```
/* Traffic Light using Red, Green and Yellow LEDs
all controlled by digital pins on the UNO32 */

const int Red = 10;
const int Yellow = 11;
const int Green = 12;

void setup() {
 pinMode(Red, OUTPUT);
 pinMode(Yellow, OUTPUT);
 pinMode(Green, OUTPUT);

}

void loop() {
  digitalWrite(Red, HIGH);      // set the LED on
  digitalWrite(Yellow, LOW);    // set the LED off
  digitalWrite(Green, LOW);     // set the LED off
  delay(2000);            // wait for two second

  digitalWrite(Red, LOW);       // set the LED off
  digitalWrite(Yellow, LOW);    // set the LED off
  digitalWrite(Green, HIGH);    // set the LED on
  delay(2000);            // wait for two second

  digitalWrite(Red, LOW);       // set the LED off
  digitalWrite(Yellow, HIGH);   // set the LED on
  digitalWrite(Green, LOW);     // set the LED off
  delay(500);    // Wait for half second
}
```

How It Works

The first section creates labels for the digital pins by using the "const" directive which is short for constant and the "int" directive which is short for integer. The label "Red" is associated with the digital pin number 10. It cannot change while running the sketch since the "const" makes the value constant. Anytime the label Red is used in the Sketch, the compiler with automatically replace it with the number 10 before it produces the code that gets sent to the UNO32.

By creating a label for each digital pin that matches the LED color makes it easier to understand which traffic light color is being controlled. These constants are created before we enter the setup loop. They are only used by the compiler to build the sketch and not part of the code that is run in the UNO32.

```
const int Red = 10;
const int Yellow = 11;
const int Green = 12;
```

The next section is the setup loop that runs one time in the UNO32. Here we use the pinmode function to make each digital pin an output so they can control the LED. A high on the digital pin will light the LED and a low will shut it off.

```
void setup() {
 pinMode(Red, OUTPUT);
 pinMode(Yellow, OUTPUT);
```

```
  pinMode(Green, OUTPUT);
}
```

The main loop is where the traffic light comes to life. There are three sections. The first lights the red LED while making sure the yellow and green are off. Then the green is lit next followed by the yellow. The digitalWrite function controls the digital pins and a high in the command is what lights the LED. You can see only three lines have the high in the command line.

There are delays after each setting of the LEDs. The red and green stay lit for two seconds by using a delay(2000) function. The yellow is lit for only a half of a second with the delay(500) line.

```
void loop() {
  digitalWrite(Red, HIGH);      // set the LED on
  digitalWrite(Yellow, LOW);    // set the LED off
  digitalWrite(Green, LOW);     // set the LED off
  delay(2000);            // wait for two second

  digitalWrite(Red, LOW);       // set the LED off
  digitalWrite(Yellow, LOW);    // set the LED off
  digitalWrite(Green, HIGH);    // set the LED on
  delay(2000);            // wait for two second

  digitalWrite(Red, LOW);       // set the LED off
  digitalWrite(Yellow, HIGH);   // set the LED on
  digitalWrite(Green, LOW);     // set the LED off
  delay(500);    // Wait for half second
}
```

As you can see this is a very short sketch but is doing something useful.

Next Steps

Obvious next steps would be to add more LEDs to create a four sided traffic light. This will take a little more software but should be easy to do. If you have electronics experience then you can have the digital pins drive a transistor or even a relay so you can control a higher current light such as a light bulb or high brightness LED. Then you could actually make a real traffic light that is visible outside. That might be a fun project to build for children in school just learning about crossing the street.

Chapter 5 – Scroll LEDs

A popular TV show from the 80's was Night Rider which featured a computer controlled car that had lights that scrolled back and forth on the front of the car. That effect is easy to produce with the UNO32. For this project I'll introduce the For Loop function. This makes the program a little shorter than the software method used in the traffic light project which had a separate digitalWrite function for each LED. The For Loop allows us to reuse a single set of digitalWrite functions by making the pin parameter into a variable.

Hardware

The hardware builds off the previous project by just adding the an extra LED. The unique operation is in the software. The four LEDs are all individually controlled by a digital I/O pin. Figure 5-1 shows the connections on the breadboard. Each LED has its own 1k resistor to limit current.

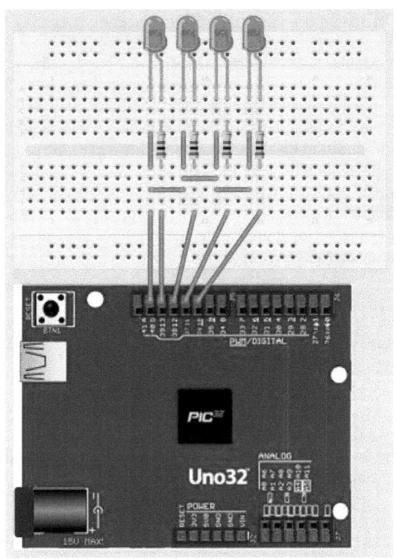

Figure 5-1: Scrolling LEDs Circuit

Software

```
/*
Scroll LEDs
Demonstrates the use of a for() loop.
Lights multiple LEDs in sequence, then in
reverse.
*/

int timer = 100; // Scroll speed control

void setup() {
// for loop to initialize each pin as an output:
  for (int LED = 10; LED < 14; LED++)  {
    pinMode(LED, OUTPUT);
  }
}

void loop() {
  // loop from the lowest pin to the highest:
  for (int LED = 10; LED <= 13; LED++) {
     // turn the pin on:
    digitalWrite(LED, HIGH);
    delay(timer);

     // turn the pin off:
    digitalWrite(LED, LOW);
  }

  // loop from the highest pin to the lowest:
  for (int LED = 13; LED >= 10; LED--) {
     // turn the pin on:
    digitalWrite(LED, HIGH);
    delay(timer);

     // turn the pin off:
    digitalWrite(LED, LOW);
  }
}
```

How it Works

After the title comment block a variable is created as a 16 bit integer (0-65535 decimal value) and preset to 100. This will be used in all delay functions so changing this one location will change the speed of the scrolling LEDs.

```
/*
Scroll LEDs
Demonstrates the use of a for() loop.
Lights multiple LEDs in sequence, then in
reverse.
*/

int timer = 100; // Scroll speed control
```

The digital pins connected to the LEDs all need to be set to outputs in the setup loop. Rather than do this four times, we use a For Loop to do it once using the variable LED and repeat it four times.

```
void loop() {
  // loop from the lowest pin to the highest:
  for (int LED = 10; LED <= 13; LED++) {
    pinMode(LED, OUTPUT);
  }
}
```

The For Loop creates the variable LED as an integer and presets it to 10. Then the second section tests the variable LED to see if it's less than or equal to 13. If it is greater than 13 then the For Loop is exited and the next block of code is run.

The last item in the For Loop is the increment equation. It is the same as writing LED = LED + 1. This is performed after every loop.

All the functions between the two curly brackets after the For statement line are part of the For Loop that get run over and over as long as the middle parameter determines that LED is less than 14.

Now we use a second For Loop to drive the LEDs high or low. In this For Loop we test if the value is less than or equal to 13. The result is the same as the first For Loop but a slightly different equation. I just wanted to show two types of equations.

```
void loop() {
  // loop from the lowest pin to the highest:
  for (int LED = 10; LED <= 13; LED++) {
     // turn the pin on:
    digitalWrite(LED, HIGH);

    delay(timer);

     // turn the pin off:
    digitalWrite(LED, LOW);
  }
```

Notice that the digitalWrite operates on the pin number stored in the variable LED. In our example that can be 10, 11, 12 or 13 which are the pins that control the LEDs.

The first digitalWrite drives the pin high to light the LED and the second drives the pin low to shut down the LED.

In between a delay function controls how long the LED is on and thus the scroll speed. This is also a variable that we created at the beginning and preset to 100.

```
// loop from the highest pin to the lowest:
for (int LED = 13; LED >= 10; LED--) {
    // turn the pin on:
  digitalWrite(LED, HIGH);
  delay(timer);

    // turn the pin off:
  digitalWrite(LED, LOW);
  }
}
```

The last section of the sketch repeats the For Loop control of the LEDs but this time the LED++ is replaced with LED- - in the For Loop. This is the same as LED = LED - 1. This makes the value of the variable LED decrement. We also start the For Loop at a higher value and test for a lower value. The LEDs will light in reverse in this For Loop thus creating the second part of the back and forth movement of light.

Next Steps

The logical next step is to change the timer variable value to speed up or slow down the scroll effect. Another option is to add more LEDs. This will require modification to the start and stop values in the For Loop but shouldn't make the sketch any larger.

Chapter 6 – Sensing a Switch

On many projects you will need some kind of human interface to control the operation. A momentary push button switch is a very common way to do that. It can start and stop the operation or it could speed up or slow down what the microcontroller is controlling. In order to do that though the software needs to recognize that a switch was pressed. This project shows a simple method of sensing a momentary push button switch.

The software will have to monitor the switch continuously as part of the main loop of code and then respond. In this project the software will light one LED until the switch is pressed at which point the LED will shut off and a second LED with light. As long as the switch is pressed the first LED will stay off and the second LED will stay on. As soon as the switch is released the first LED will once again light up and the second LED will shut off. The completed project is shown in Figure 6-1.

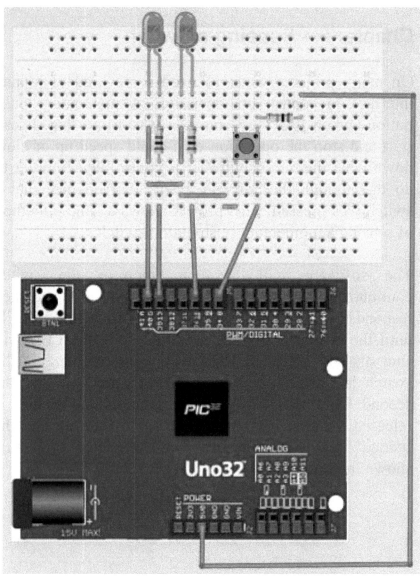

Figure 6-1: Final Switch Sensing Project

Hardware

The hardware uses the same two Red LED connections as the train crossing project in Chapter 3. The addition of the switch is shown in Figure 4-1. The switch is wired as a low side switch meaning the circuit has a pull-up resistor to 5 volts so the input the micro is high when the switch is not pressed and low when the switch is pressed. This is known as a low side switch. If the parts were reversed and the switch was connected to 5 volts and the resistor to ground then it would be a high side switch.

Note: The UNO32 has 5v tolerant inputs on the digital pins. Even though the expected voltage is 3.3v max, 5v will work without damage because the PIC32 microcontroller has internal circuitry to protect from the higher voltage on these pins. The Analog pins have protection circuitry on the UNO32 board to protect those pins. This was done to make UNO32 compatible with Arduino shields.

The software will change pin 8 to an input and then test pin 8 to see if it changes to low indicating the switch has been pressed. The LEDs are connected to pin 10 and pin 13 through 1k resistors.

Software

```
/*
 Switch

One LED is on and one off when switch is idle
and the reverse happens when the switch is
pressed.
*/
```

```
//Constants
// the number of the switch pin
const int switchPin = 8;
// the number of the LED1 pin
const int led1Pin =  10;
// the number of the LED2 pin
const int led2Pin =  13;

// Variables
// variable for reading the pushbutton status
int switchState = 0;

void setup() {
  // initialize the LED pins as outputs:
  pinMode(led1Pin, OUTPUT);
  pinMode(led2Pin, OUTPUT);

  // initialize the switch pin as an input:
  pinMode(switchPin, INPUT);
}

void loop(){
  // read the state of the switch:
 switchState = digitalRead(switchPin);

  // check if the switch is pressed.
  // if it is, the switchState is low:
  if (switchState == LOW) {
    // Switch Pressed:
    digitalWrite(led2Pin, HIGH);
    digitalWrite(led1Pin, LOW);
  }
  else {
    // Switch Idle:
    digitalWrite(led1Pin, HIGH);
    digitalWrite(led2Pin, LOW);
  }
}
```

How it Works

The first section is the header that describes the sketch

```
/*
 Switch

One LED is on and one off when switch is idle
and the reverse happens when the switch is
pressed.
*/
```

Constants are created to make the LED and Switch connections easier to follow.

```
//Constants
// the number of the switch pin
const int switchPin = 8;
// the number of the LED1 pin
const int led1Pin =  10;
// the number of the LED2 pin
const int led2Pin =  13;
```

A variable is created that will store the state of the switch reading.

```
// Variables
// variable for reading the pushbutton status
int switchState = 0;
```

Now we enter the setup function that runs one time. In it we set the LED pins to outputs and the switch pin to an input. These are all digital pin connections.

```
void setup() {
  // initialize the LED pins as outputs:
```

```
  pinMode(led1Pin, OUTPUT);
  pinMode(led2Pin, OUTPUT);

  // initialize the switch pin as an input:
  pinMode(switchPin, INPUT);
}
```

The main loop is where it all comes together. The first thing we do is read the state of the switch with a digitalRead function. If the switch is idle the variable switchPin will contain a 1 value (or HIGH). If the switch is pressed then the switchPin variable will contain a 0 value (or LOW).

```
void loop(){
  // read the state of the switch:
 switchState = digitalRead(switchPin);
```

I will use an If Else statement to respond to the state of the switch. The If Else has two sections. If the switchState variable matches the If statement setting, in this case LOW or 0, then we know the switch was pressed and everything between the first set of curly brackets is executed. Two digitalWrite functions set the LEDs.

```
  // check if the switch is pressed.
  // if it is, the switchState is low:

  if (switchState == LOW) {
    // Switch Pressed:
    digitalWrite(led2Pin, HIGH);
    digitalWrite(led1Pin, LOW);
  }
```

If instead the switchState variable is idle then the section below "else" is executed which reverses the LEDs.

```
else {
  // Switch Idle:
  digitalWrite(led1Pin, HIGH);
  digitalWrite(led2Pin, LOW);
}
}
```

Next Steps

You could add a second switch so one turns the LED on and the second turns it off. Another option is to make an LED toggle or turn on with one press and then off on a second press. You could also use the scrolling LED project setup of Chapter 5 but instead of it automatically scrolling through the LEDs, let the switch control it.

Chapter 7 - Read a Potentiometer

Many sensors output a signal that is actually a variable voltage also known as an analog voltage. To convert that voltage into a digital value the UNO32 uses the Analog to Digital Converter (ADC) built into the PIC32 microcontroller. The ADC pins are on the analog header. The UNO32 software has a specific function for reading analog pins and storing the measurement in a variable as a digital value. This makes reading a sensor voltage as easy as a few lines of code in a sketch. In this Chapter's project we will create our own adjustable voltage with a potentiometer connected to the A0 pin. Based on the position of the potentiometer we will light up different LEDs similar to the volume display on a stereo system. The completed project is shown in Figure 7-1.

Hardware

The hardware is similar to other projects with the LEDs wired to four separate pins. The potentiometer is a 10k value that is wired with 3.3v on one side and ground on the other. The center wiper connection is connected to the A0 pin of the UNO32. You can also connect the potentiometer to 5v as the UNO32 has protection on the analog pins to prevent damage for voltages up to 5v. The PIC32 is limited to 3.3v input sensing though so parts of the potentiometer travel will not be sensed once the voltage reaches 3.3v when connected to 5v. I mention this because the demo shield mentioned earlier that makes the electrical connections easier has the potentiometer connected to 5v.

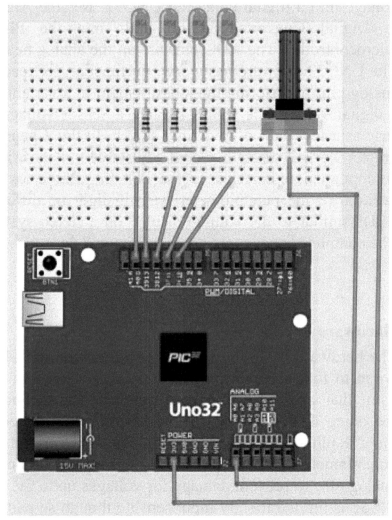

Figure 7-1: Final Potentiometer Project

Software

```
/*
ADC
Reads an analog input on pin A0, and lights four
LEDs accordingly.
*/

void setup() {
 pinMode(10, OUTPUT);
 pinMode(11, OUTPUT);
 pinMode(12, OUTPUT);
 pinMode(13, OUTPUT);

}

void loop() {
  int sensor = analogRead(0);

  if (sensor < 150) {
  digitalWrite(10, LOW);   // set the LED off
  digitalWrite(11, LOW);    // set the LED off
  digitalWrite(12, LOW);    // set the LED off
  digitalWrite(13, LOW);    // set the LED off
  }

  if (sensor > 150) {
  digitalWrite(10, HIGH);   // set the LED on
  digitalWrite(11, LOW);    // set the LED off
  digitalWrite(12, LOW);    // set the LED off
  digitalWrite(13, LOW);    // set the LED off
  }

  if (sensor > 500) {
  digitalWrite(10, HIGH);   // set the LED on
  digitalWrite(11, HIGH);    // set the LED on
  digitalWrite(12, LOW);    // set the LED off
  digitalWrite(13, LOW);    // set the LED off
  }
```

```
if (sensor > 750) {
digitalWrite(10, HIGH);   // set the LED on
digitalWrite(11, HIGH);    // set the LED on
digitalWrite(12, HIGH);   // set the LED on
digitalWrite(13, LOW);     // set the LED off
}

if (sensor > 1000) {
digitalWrite(10, HIGH);   // set the LED on
digitalWrite(11, HIGH);    // set the LED on
digitalWrite(12, HIGH);   // set the LED on
digitalWrite(13, HIGH);    // set the LED on
}

}
```

How it Works

The software sketch is longer than the previous projects but you'll soon see that it's just filled with repeating routines. It's really not that complicated.

The first section includes the header that describes the project.

```
/*
ADC
Reads an analog input on pin A0, and lights four
LEDs accordingly.
*/
```

The setup section is next and simply creates four outputs to drive the LEDs. Pins 10-13 are used.

```
void setup() {
 pinMode(10, OUTPUT);
```

```
pinMode(11, OUTPUT);
pinMode(12, OUTPUT);
pinMode(13, OUTPUT);

}
```

The main loop contains the section that actually reads the sensor and it's only one line long. The variable "sensor" is created as an int or integer. Therefore it can hold a value of 0 to 65,535 decimal. The ADC in the UNO32 is only 10 bits wide so the digital value created by the ADC will range from 0 to 1024. Zero is when the voltage is at ground and 1024 when the voltage is at 3.3v or above.

The variable "sensor" is then made equal to the result of the analogRead( ) function on pin A0. This function senses the voltage and converts it to 0-1024 and stores it in the variable sensor.

```
void loop() {
  int sensor = analogRead(0);
```

Now that we have the converted value store in sensor we can test that variable to determine which LEDs to light. Through five different IF statements we create the moving light as the potentiometer is turned.

The first one tests if the value of sensor is less than 150. If the value is below 150 then all the LEDs are off by setting the pins low.

```
  if (sensor < 150) {
```

```
digitalWrite(10, LOW);    // set the LED off
digitalWrite(11, LOW);    // set the LED off
digitalWrite(12, LOW);    // set the LED off
digitalWrite(13, LOW);    // set the LED off
}
```

The next section tests if the sensor value is larger than 150. If it is the first LED is lit.

```
if (sensor > 150) {
  digitalWrite(10, HIGH);   // set the LED on
  digitalWrite(11, LOW);    // set the LED off
  digitalWrite(12, LOW);    // set the LED off
  digitalWrite(13, LOW);    // set the LED off
}
```

The next three section test the value against 500, 750 and 1000 using a separate IF statement. Each IF statement lights a different number of LEDs. The program runs through each IF statement to see which one matches and the last one to match is how the LEDs will finally be lit.

The main loop reads the analog value on each loop through. As the potentiometer is changed, the sensor variable will also change and the LEDs will match based on the IF statements.

Next Steps

A simple next step is to change the values of the IF statement lines to change when to light up the LEDs. You could also change the LED arrangement to light the LEDs differently. You could easily add more IF statement lines and connect more LEDs. This would allow you to indicate more of the potentiometer movement.

Chapter 8 - Sensing Light

Similar to the previous project, we can use a photo resistor to sense light using an analog pin. A photo resistor changes resistance as it is exposed to light. By putting a pull-up resistor connected to 3.3v in series with the photoresistor, changing light will create a variable voltage. The variable voltage can then be used similar to the way we read the potentiometer. In this case though we'll just light an LED when it's dark and turn it off when there is light.

Hardware

The layout of the light sensor and pull-up resistor are very similar to the potentiometer in the last project. One side of the light sensor is connected to ground. The other connects to the resistor and also the analog pin 1. The pull-up resistor is a 10k connected to 3.3v. You may have to adjust the resistor for your light sensor but in most cases a 10k will work fine. In the dark the photo resistor has a high resistance and in the light it has a low resistance. Therefore in the dark the voltage will be high and in the light the voltage will be low.

The LED controlled by the sensor is connected to pin 10 of the digital port. A 220 ohm series resistor connected to the anode and the cathode connected to ground completes the circuit.

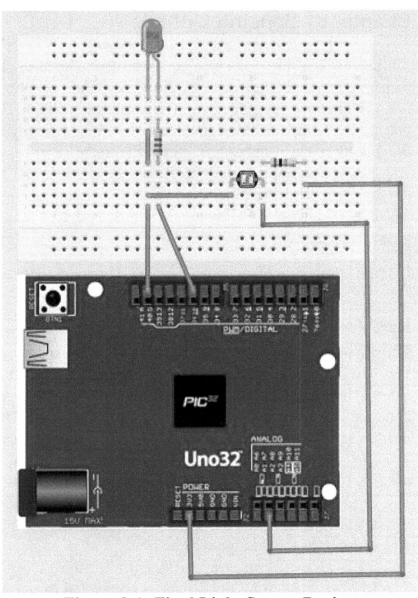

Figure 8-1: Final Light Sensor Project

Software

```
/*
Light
Reads the voltage on pin A1 connected to a light
sensor.
When in dark the LED lights.
*/


void setup() {
 pinMode(10, OUTPUT);
}

void loop() {
  int sensor = analogRead(1);

  if (sensor < 500) {
  digitalWrite(10, LOW);    // LED off
  }
  else {
  digitalWrite(10, HIGH);   // LED on
  }

}
```

How it Works

As usual the header block starts the program. This is optional but makes it easier to read the program and know what it does.

```
/*
Light
Reads the voltage on pin A1 connected to a light
sensor. When in dark the LED lights.
*/
```

The setup loop makes the pin connected to the LED an output. We don't have to touch the analog pin as the setting for that pin is preset to an analog input.

```
void setup() {
 pinMode(10, OUTPUT);
}
```

The main loop is where the magic happens. The analogRead function converts the light sensor voltage to a digital value and stores it in the integer variable "sensor".

```
void loop() {
   int sensor = analogRead(1);
```

In this project we use the IF ELSE statement. This gives us two choices for each IF test. If the value of the variable sensor is below 500 then it has light on it. This equates to around 1.6 volts. The LED is set to off by making pin 10 low. If instead the IF statement is false and the value is greater than or equal to 500 then the ELSE portion is executed and pin 10 is set high. This lights the LED which should be lighting in the dark.

```
   if (sensor < 500) {
   digitalWrite(10, LOW);    // LED off
   }
   else {
   digitalWrite(10, HIGH);   // LED on
   }

}
```

Next Steps

Changing the threshold value is an obvious option. You could combine the previous potentiometer project with this and use the potentiometer to set the test value in the light sensor project. This way you could adjust the sensitivity of the photoresistor by adjusting the potentiometer. The demo shield has these connected separately to A0 and A1 so you can do this without any hardware changes.

Chapter 9 – Creating Sound

Creating sound through a speaker can be very useful for a lot of projects. You can create a sound when a switch is pressed or sound an alarm when the temperature drops. In this Chapter I'll show you how to make a tone through a speaker using a digital pin. I've selected a PWM pin because the demo-shield is wired like that but this can be run on any digital pin. The Tone function in the chipKIT library will be used to make this a very short program to write. This project will create the sound of a falling object similar to what you might hear on an old video game.

Hardware

The UNO32 will produce a square wave because it's a digital pin driving the speaker. The square wave can be converted into a semi-rounded signal to work better with the speaker by placing a 10 uf capacitor in series between pin 5 and the speakers positive lead. The other side of the speaker is then grounded. An 8-ohm speaker works fine but a more common approach is a piezo speaker. A Piezo speaker from Jameco.com under part number DBX05-PN will work well.

The setup is shown in Figure 9-1. The right side of the capacitor is the positive side and that connects to the digital pin of the UNO32. The speaker positive is also on the right side and connects to the negative side of the capacitor. The speaker is then grounded to the ground pin on the UNO32 header.

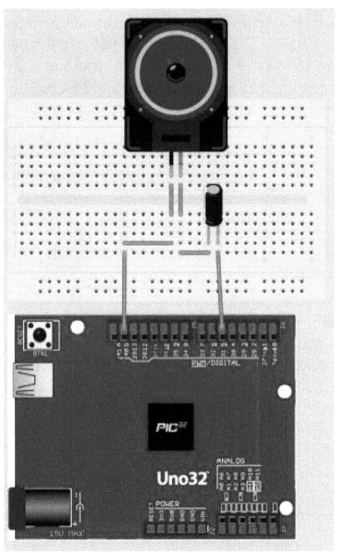

Figure 9-1: Final Creating Sound Project

Software

```
/* Sound
Create the sound of a falling object.
*/

void setup() {
}

void loop() {
  for (int x = 2000; x>200; x=x-5)
  {
    tone(5, x, 100);   //Output Tone
    delay(10);         //Delay between tones
  }

    noTone(5);         //Tone off
    delay(1000);       //Wait 1 second and repeat
}
```

How it Works

The first section includes the description header and then the setup section. In this project we don't need to setup the pins as the Tone function handles that setup.

```
/* Sound
Create the sound of a falling object.
*/

void setup() {
}
```

The loop begins with a FOR loop that starts by setting an integer variable "x" to a high value of 2000 and continues until the variable x is less than 200. The value of x is decremented by 5 every loop. The value of x will be used as the frequency setting in the Tone function.

```
void loop() {
    for (int x = 2000; x>200; x=x-5)
    {
```

The Tone function has three parameters. The pin number is the first parameter and is the pin that the signal is sent from. The frequency of the tone is next and the duration of the tone in milliseconds is the final parameter. In this case pin 5 is used as the output pin. The frequency is set by the value of the variable x which we set to 2000 initially and drop to 200 over time. This creates the falling sound. Each note is played for 100 milliseconds. Then a 10 millisecond delay is executed before fetching the next tone.

```
    tone(5, x, 100);    //Output Tone
    delay(10);          //Delay between tones
}
```

The tone is then stopped with a noTone function. This creates a silence. The silence lasts for 1 second using a delay function. This completes the loop and will repeat every second until you are so annoyed you pull the power. At least that's what I did.

```
    noTone(5);          //Tone off
    delay(1000);        //Wait 1 second and repeat
}
```

Next Steps

There are many options to this simple sketch. You can make the piezo sound last longer or shorter by changing the values in the Tone function. You can change the range of the variable "x" or even have it increase instead of decrease. You can use the routine with the light sensor or potentiometer to change the value of the x based on those sensors. You can also add a switch that the software monitors so you can turn it on or off. I highly recommend this option!

Chapter 10 – Dimming an LED with PWM

The term PWM stands for Pulse Width Modulation which is just a way to control how long something is on vs off or high vs low. The amount of time it is high is called the duty cycle. The PWM signal will have a constant frequency and a variable duty cycle. A 50% duty cycle will be half the time high and half the time low on each cycle. A 25% duty cycle will be high ¼ of the time and ¾ if the time low. The longer the duty cycle the higher the average voltage produced by the signal.

If we take that signal and drive an LED, the LED will be brighter when the duty cycle is higher than when it's lower. A 100% duty cycle is high all the time and will continuously light the LED. A 50% will drop it to about half the brightness. A 0% duty cycle has the LED off all the time. A PWM signal will typically be less than 100% and greater than 0%. By changing the duty cycle we can change the brightness of the LED. This project will do just that in a continuous loop.

Hardware

The hardware is the same setup as driving an LED in the blink project of Chapter 2. The difference is software. The LED is connected to pin 10 of the UNO32 which is a PWM pin. This is designated by the line under the pin number on the board. The LED is then driven by pin 10.

The anode (longer pin) of the LED is connected to pin 10 through a 220 ohm resistor. The LED is then grounded by connecting the cathode to the ground pin of the UNO32 header.

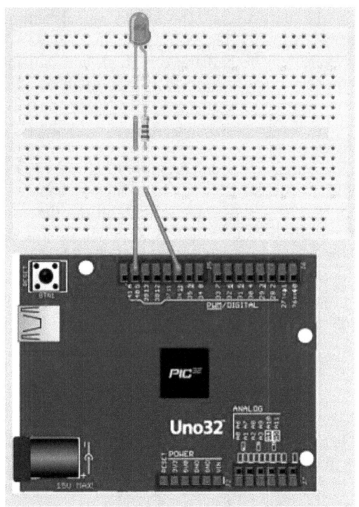

Figure 10-1: Final Sensing Vibration Project

Software

```
/*
 PWM
 This example shows how to pulse an LED using the
 analogWrite() function.
 */

int ledPin = 10;   //LED connected to PWM pin 10

void setup()   {
  // nothing happens in setup
}

void loop()   {
  // Change from Dim to Bright
  for(int dim = 0 ; dim <= 255; dim=dim+5)
  {
    analogWrite(ledPin, dim);
    delay(20);
  }

  // Change from Bright to Dim
  for(int dim = 255 ; dim >= 0; dim=dim-5)
  {
    analogWrite(ledPin, dim);
    delay(20);
  }
}
```

How it Works

The first section describes the project in the header and then creates a nickname for the LED PWM pin.

```
/*
 PWM
 This example shows how to pulse an LED using the
 analogWrite() function.
 */

int ledPin = 10;   //LED connected to PWM pin 10
```

The setup section doesn't require any actions since the analogWrite function handles the settings of the pin.

```
void setup()   {
  // nothing happens in setup
}
```

The main loop first uses a FOR loop to create an increasing variable named "dim". It changes from 0 to 255 in increments of 5 counts.

```
void loop()   {
  // Change from Dim to Bright
  for(int dim = 0 ; dim <= 255; dim=dim+5)
  {
```

The analogWrite function does all the work. The pin is the first parameter and the variable "dim" is the second. The value of "dim" controls the duty cycle. A value of 0 equals a duty cycle of 0. A value of 255 equals a duty cycle of 100%. Therefore this section drives the LED from off to fully on. A delay of 20 milliseconds is added after to allow us to see the LED at each brightness level.

```
  analogWrite(ledPin, dim);
  delay(20);
}
```

The next section does the opposite and creates a FOR
loop that decreases the variable dim from 255 down to 0.
This is then used in the analogWrite function to drive the
LED from 100% duty cycle to 0% duty cycle. This dims
the LED. Another delay is added.

```
// Change from Bright to Dim
for(int dim = 255 ; dim >= 0; dim=dim-5)
{
  analogWrite(ledPin, dim);
  delay(20);
}
}
```

This brightening and dimming of the LED continues in a
loop creating a heart beat kind of effect in the LED.

Next Steps

The first thing I would try is to change the step size to see
if you can get it to be smoother or jumpier depending on
what effect you want. You could even change the delay
time to make the LED dim slower or brighten faster.
You could also combine this with the light sensor project
and dim the LED to match the outside brightness. To do

this just use the analogRead function to read the light sensor and adjust the dim variable. You won't need the For loops, just the analogWrite lines.

Chapter 11 – Serial Communication

In any electronic project, having a way to monitor a variable or calculation in a program is a great debug tool. In the UNO32 you can easily setup a communication channel using the USB connection. The chipKIT MPIDE software has a built in terminal screen that will display any information sent serially through the USB connection. In this project we will once again read a potentiometer with the analogRead function but instead of driving LEDs, we will just send the measured value back to the PC through the USB connection. The terminal screen will then display the value on the PC.

Hardware

The hardware setup is simple which only requires a potentiometer and three connections. The potentiometer has 3.3v connected to one side and ground to the other. The center tap of the potentiometer is connected to the A0 pin so we can read the voltage with the analogRead function. The serial communication hardware is built into the UNO32 so we don't need anything added for that. The setup is shown in Figure 11-1.

If you use the CHIPINO demo shield, then the potentiometer is connected to 5 volts. The potentiometer movement won't be sensed once the voltage at A0 exceeds 3.3v but it will still work fine for this project.

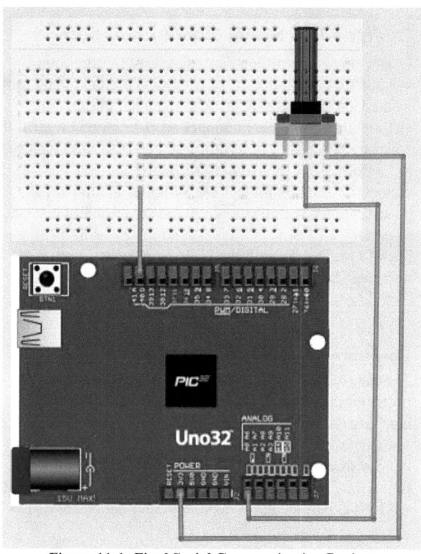

Figure 11-1: Final Serial Communication Project

Software

```
/*
 Serial
 Reads the voltage on the potentiometer and
 sends the result to the serial monitor
 */

void setup() {
  Serial.begin(9600);
}

void loop() {
  int sensor = analogRead(0);
  Serial.println(sensor, DEC);
}
```

After you program the UNO32 you launch the serial monitor by clicking on the ICON at the far right as seen in the picture below.

Figure 11-2: MPIDE Icons

The Serial Terminal will launch and begin to receive the data from the UNO32. The value of the potentiometer reading will display over and over again as seen in Figure 11-3. The lower right hand corner shows the baud rate that should match the setting in the setup section. The default is 9600.

111

Figure 11-3: Serial Monitor Screen

How it Works

The header describes the project which is actually a real short software project despite how complex this could be if chipKIT didn't have the serial function.

```
/*
Serial
Reads the voltage on the potentiometer and
sends the result to the serial monitor
*/
```

The setup section requires us to establish the serial communication speed. In this case we use the 9600 baud rate setting. The Serial.begin function is how we establish this connection.

```
void setup() {
  Serial.begin(9600);
}
```

The main loop is where the sensor is read. A variable named "sensor" is created as an integer value which can hold any value from 0 to 65,535. The ADC in the UNO32 as we learned earlier is 10-bit so it will create a digital value from 0 to 1024. This will get stored in the variable "sensor".

```
void loop() {
  int sensor = analogRead(0);
```

The serial communication is one command line long. The Serial.println function has the variable "sensor" as its parameter. It also has the modifier DEC included. This converts the binary value of sensor into the ASCII characters needed to create the decimal value of the sensor value. The println portion indicates to send a carriage return and line feed after sending the data serially. This creates a clean fresh line in the terminal window for each potentiometer reading sent.

```
  Serial.println(sensor, DEC);
}
```

The single Serial.println function does a lot to simplfy sending data to the terminal window. You can see how easy it would be to add these few lines to any project and get serial data on any variable in a sketch.

Next Steps

Try replacing the DEC modifier with HEX or BIN and see what the serial monitor shows. I think you will find it interesting. There are other modifiers as well so you could try all of them to see what they do. Once you are comfortable with this, try adding a serial line to any of the earlier projects to monitor a variable.

Conclusion

I hope you were able to gain a fundamental understanding of how to program an UNO32 module. I tried to pick projects that were easy enough to complete in a short time but gave you all the critical details you will need to build any future UNO32 projects. Knowing how to control a digital output, read a digital input, read an analog input and control an analog output will be used in some form in most the projects you create with the UNO32 module.

There are far more advanced sketch examples out there but all the same programming and steps are the same. If you get a strange result, use the serial monitor to determine where the sketch may be doing something you didn't expect. I'm sure with a little time you will be building projects far beyond the simple projects in this book.

I also have written books on C programming targeted for the beginner title Beginner's Guide to Embedded C Programming Volumes 1, 2 and 3. If you need tips on C programming used in the chipKIT environment then these books may help answer some questions.

If you have any questions or comments, please don't hesitate to send them to me at chuck@elproducts.com. I try to answer all the reader email. Thanks for buying this book. I hope you found it helpful.

Appendix A –Parts List for Projects

1 – chipKIT UNO32 Module
1 – USB Cable with mini B connector
1 – Normally Open Momentary Switch (Jameco.com #199726)
1 – Piezo Speaker DBX05-PN (Jameco.com #138740)
1 – Photoresistor (Jameco #202366)
1 – 10k Trim Potentiometer (Jameco #43001)
1 – 10uf 16v Electrolytic Capacitor (Jameco #198839)
2 – Red Diffused LED T-1 ¾ (Jameco #333973 min qty 10 pcs)
1 – Yellow Diffused LED T-1 ¾ (Jameco #333622 min qty 10 pcs)
1 – Green Diffused LED T-1 ¾ (Jameco #253833 min qty 10 pcs)
4 – 220 or 1k Ohm ¼ w Resistor (Jameco #690865 min 100 pcs)
1 – 10k ¼ w Resistor (Jameco #691104 min qty 100 pcs)

Or

1 – chipKIT UNO32 Module
1 – USB Cable with mini B connector
1 – CHIPINO Demo Shield (chipaxe.com)

Software Download
All the project software can be downloaded from my website at:

www.elproducts.com/chipkitbookfiles

Appendix B – UNO32 Pin Map

These diagrams were developed by member Doc_Norway on the chipKIT forum. I found these useful as a reference for anybody needing the connection scheme to the PIC32 chip so I included them here with their permission.

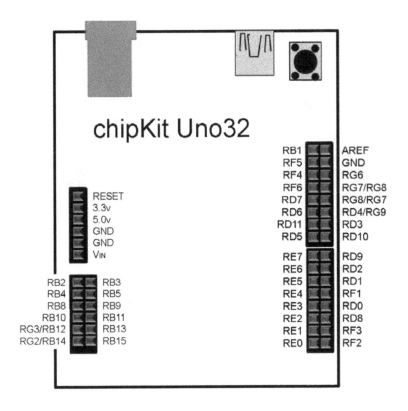

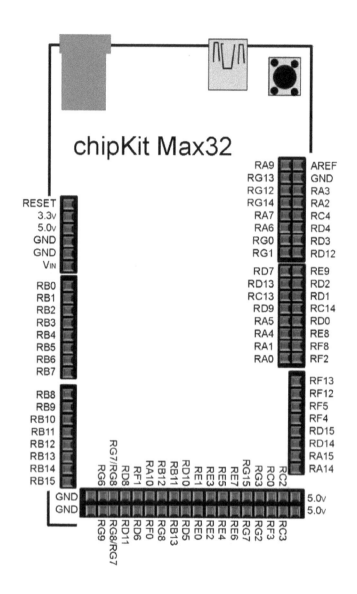

Appendix C – CHIPINO Demo Shield

CHIPINO Demo Shield

The CHIPINO shield is designed to work with any Arduino compatible module. The features of this board included everything I needed to show how to use the chipKIT module. You don't need to use this board but it does make it easier for those that are not yet comfortable with connecting the hardware on a breadboard circuit. This shield is available at chipaxe.com and more sources I'm sure after this book is released. Email me if you can't find the shield and I'll find you a source. The schematic is shown on the next page.

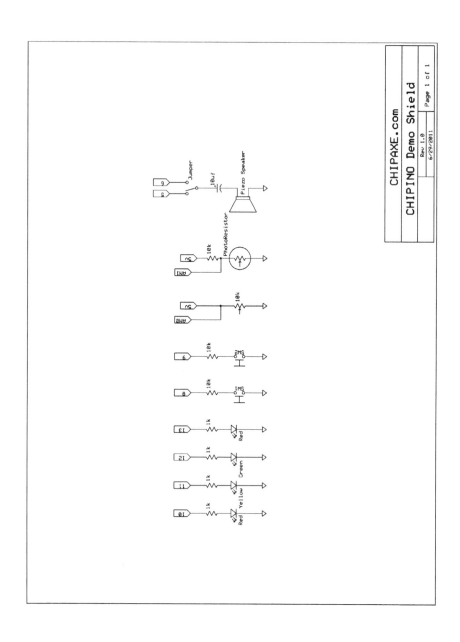

Index

Other Books by Chuck Hellebuyck:

Beginner's Guide to Embedded C Programming
Beginner's Guide to Embedded C Programming – Vol 2
Beginner's Guide to Embedded C Programming – Vol 3
Programming PICs in BASIC
Programming PIC Microcontrollers with PICBASIC
Programming the BASIC Atom Microcontroller
Getting Started with PICs – 2006
Getting Started with PICs – 2007

You can get these books at:
amazon.com
elproducts.com
nutsvolts.com
microchipdirect.com
electronics123.com

www.ingramcontent.com/pod-product-compliance
Lightning Source LLC
Chambersburg PA
CBHW071222050326
40689CB00011B/2410